# The Tree Farmer's Twelve Days of Christmas

Written by **Aaron Burakoff**

Illustrated by **Marcia Adams Ho**

On the first day of Christmas my farmer gave to me a seed to grow a new tree.

**Christmas tree seeds are found inside the cones that grow on the trees.**

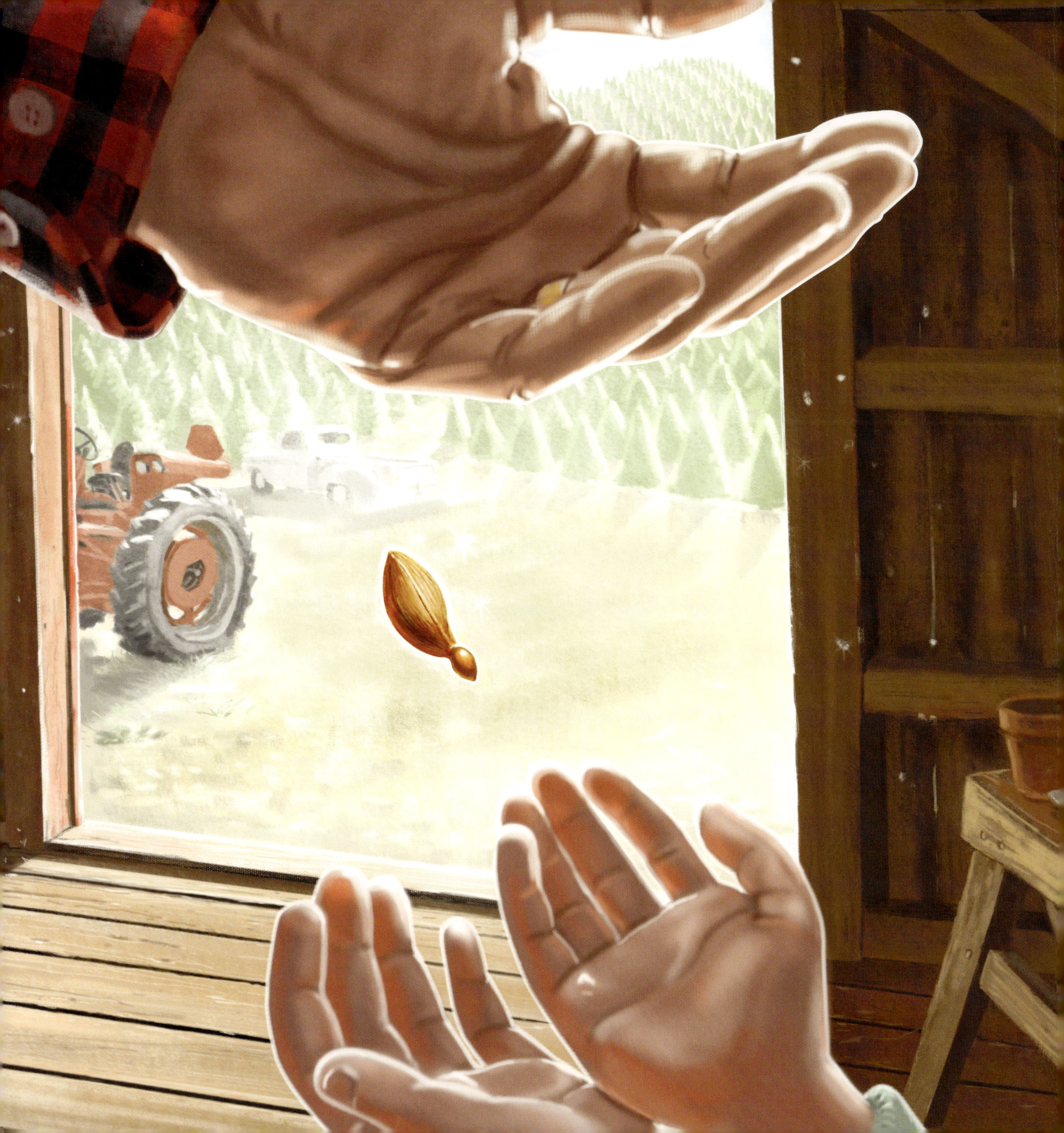

On the second day of Christmas my farmer gave to me TWO leather gloves and a seed to grow a new tree.

Tree farmers use sharp tools so they wear gloves to protect their hands.

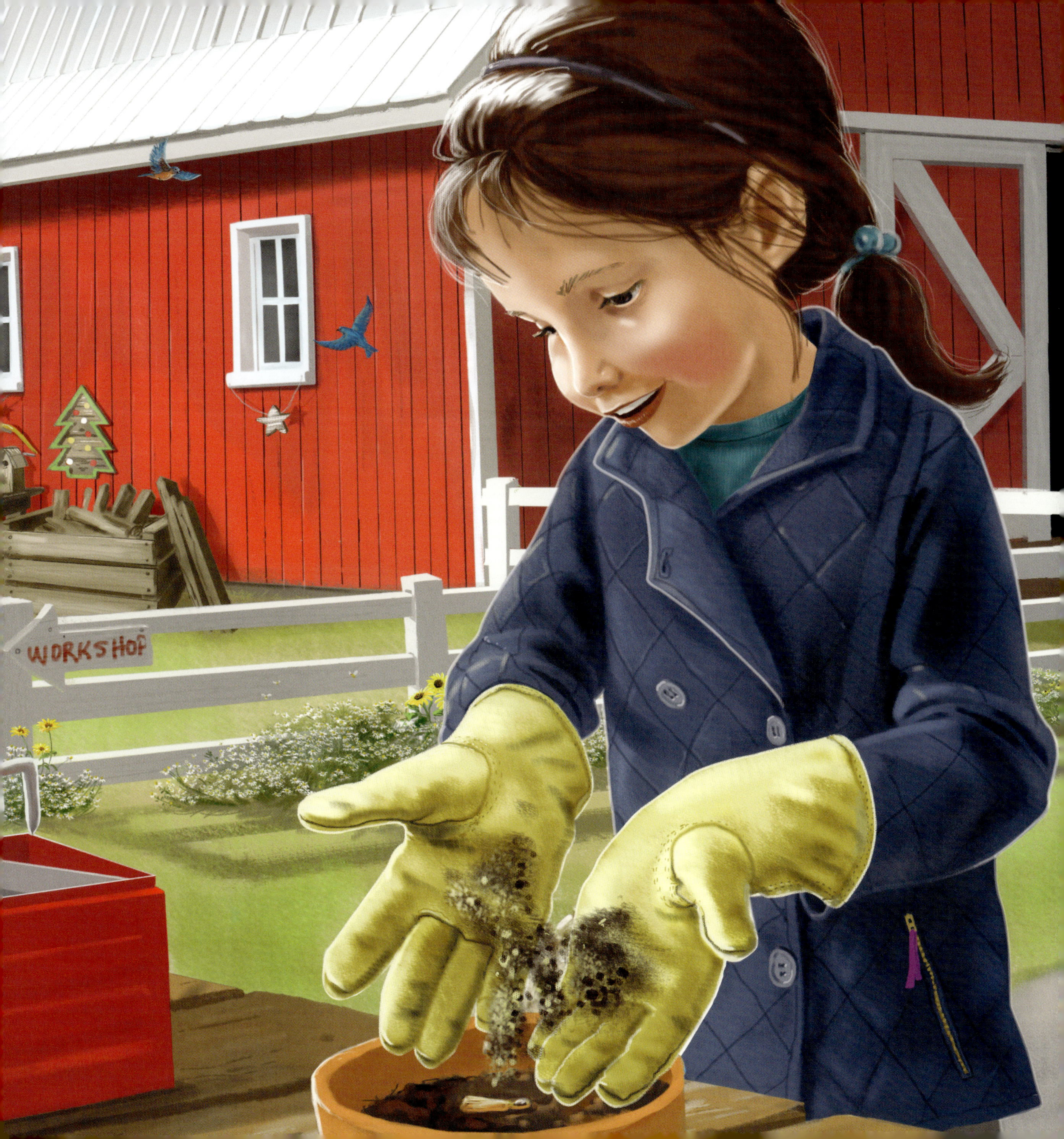
WORKSHOP

On the third day of Christmas my farmer gave to me THREE drenched friends,

Farmers must work rain or shine. Rainfall helps trees grow healthy and strong.

TWO leather gloves,
and a seed to grow a new tree.

On the fourth day of Christmas my farmer gave to me FOUR kinds of Firs,

There are many different species of Christmas trees. Some of the most common are Balsam Fir, Douglas Fir, Fraser Fir, Noble Fir, Scotch Pine, and White Pine. Ask your local farmers what kinds they grow!

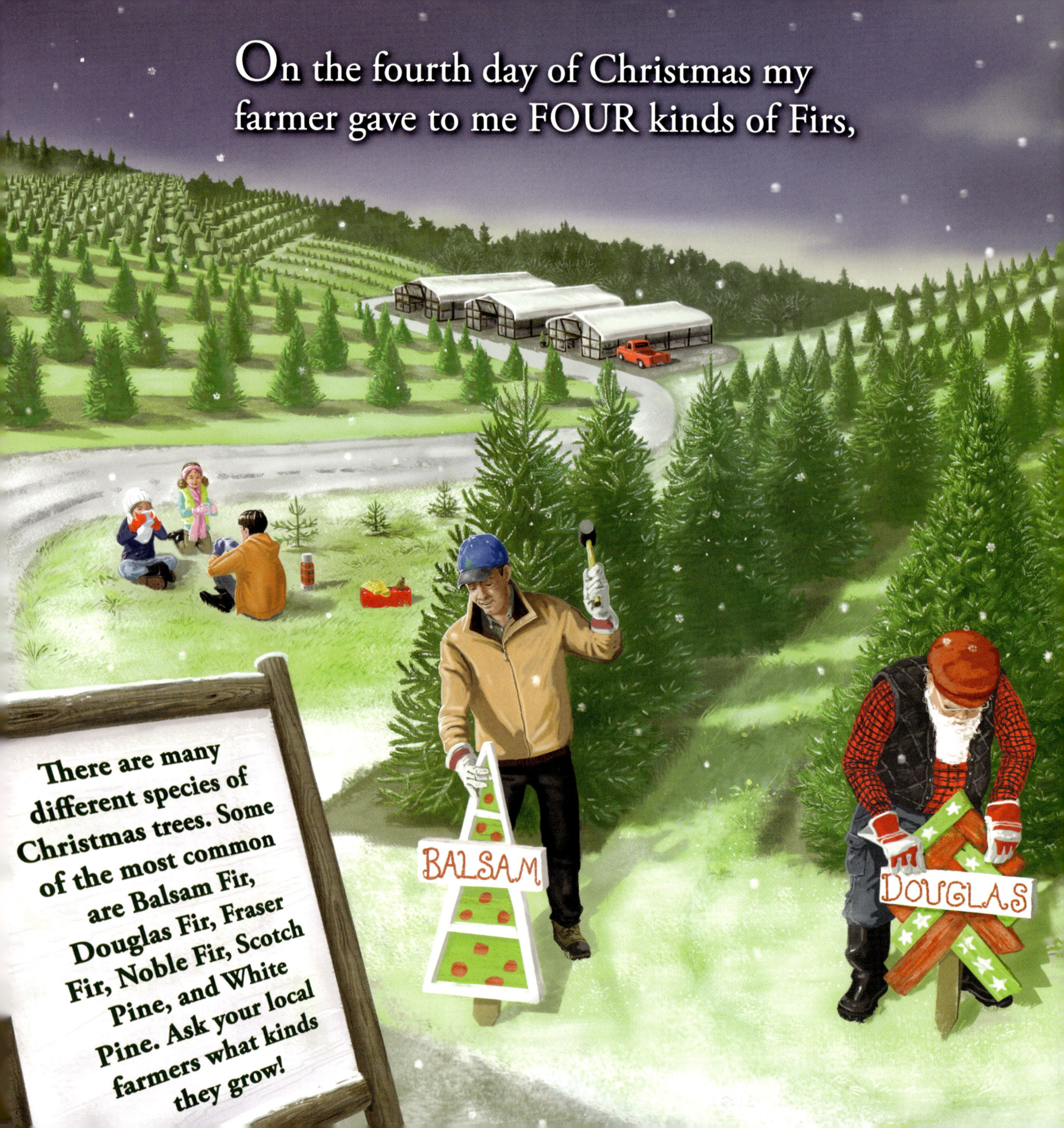

THREE drenched friends,
TWO leather gloves,
and a seed to grow a new tree.

Wreaths are popular holiday decorations made from evergreen branches decorated with pine cones.

On the fifth day of Christmas my farmer gave to me FIVE woven rings!

FOUR kinds of Firs,
THREE drenched friends,
TWO leather gloves,
and a seed to grow a new tree.

On the sixth day of Christmas my farmer gave to me SIX birds-a-nesting,

Tree farms serve as habitats for many species of birds and wildlife.

FIVE woven rings,
FOUR kinds of Firs,
THREE drenched friends,
TWO leather gloves, and
a seed to grow a new tree.

On the seventh day of Christmas
my farmer gave to me
SEVEN shearers shearing,

Farmers cut, or "shear," tree branches to maintain the trees' classic shape.

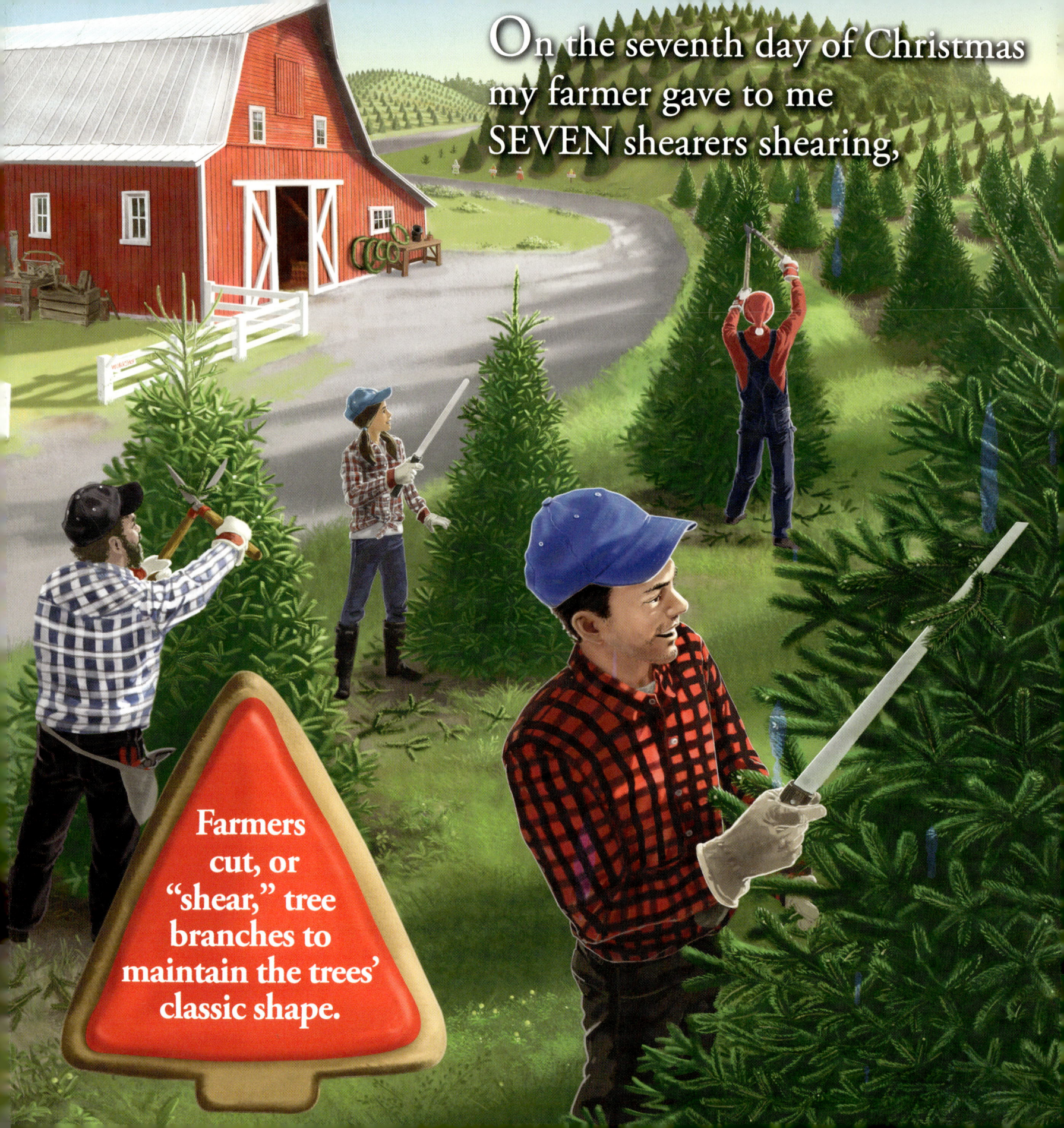

SIX birds-a-nesting, FIVE woven rings,
FOUR kinds of Firs, THREE drenched friends,
TWO leather gloves, and a seed to grow a new tree.

On the eighth day of Christmas my farmer gave to me EIGHT mowers mowing,

Farmers mow the land around the trees to remove the grasses and weeds that compete with the trees for water, sunshine, and nutrients from the soil.

SEVEN shearers shearing, SIX birds-a-nesting,
FIVE woven rings, FOUR kinds of Firs,
THREE drenched friends, TWO leather gloves,
and a seed to grow a new tree.

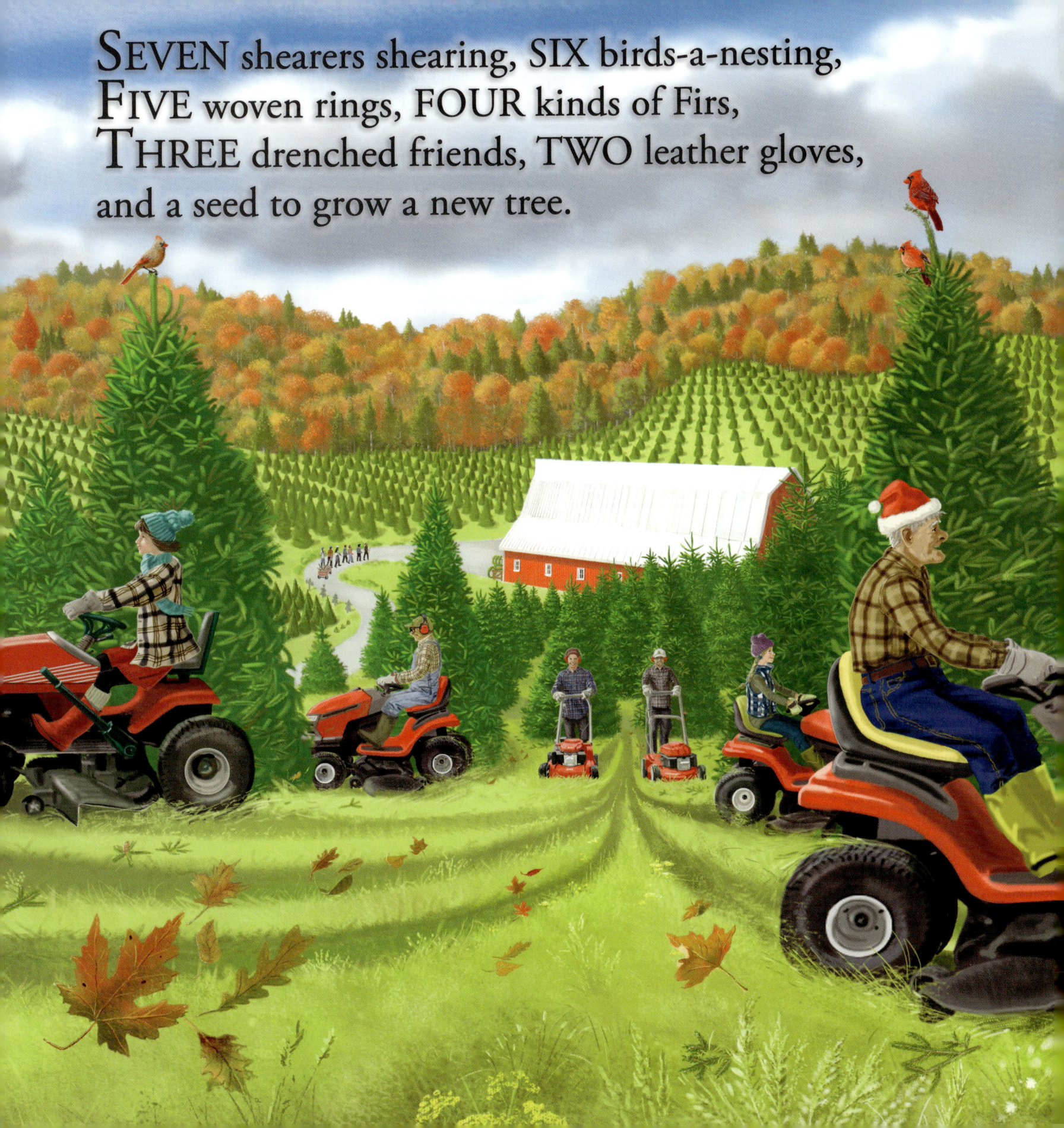

On the ninth day of Christmas my farmer gave to me NINE lanes of planting,

Farmers plant trees in well-spaced rows so the trees have enough room to grow and the farmers have enough space to care for them.

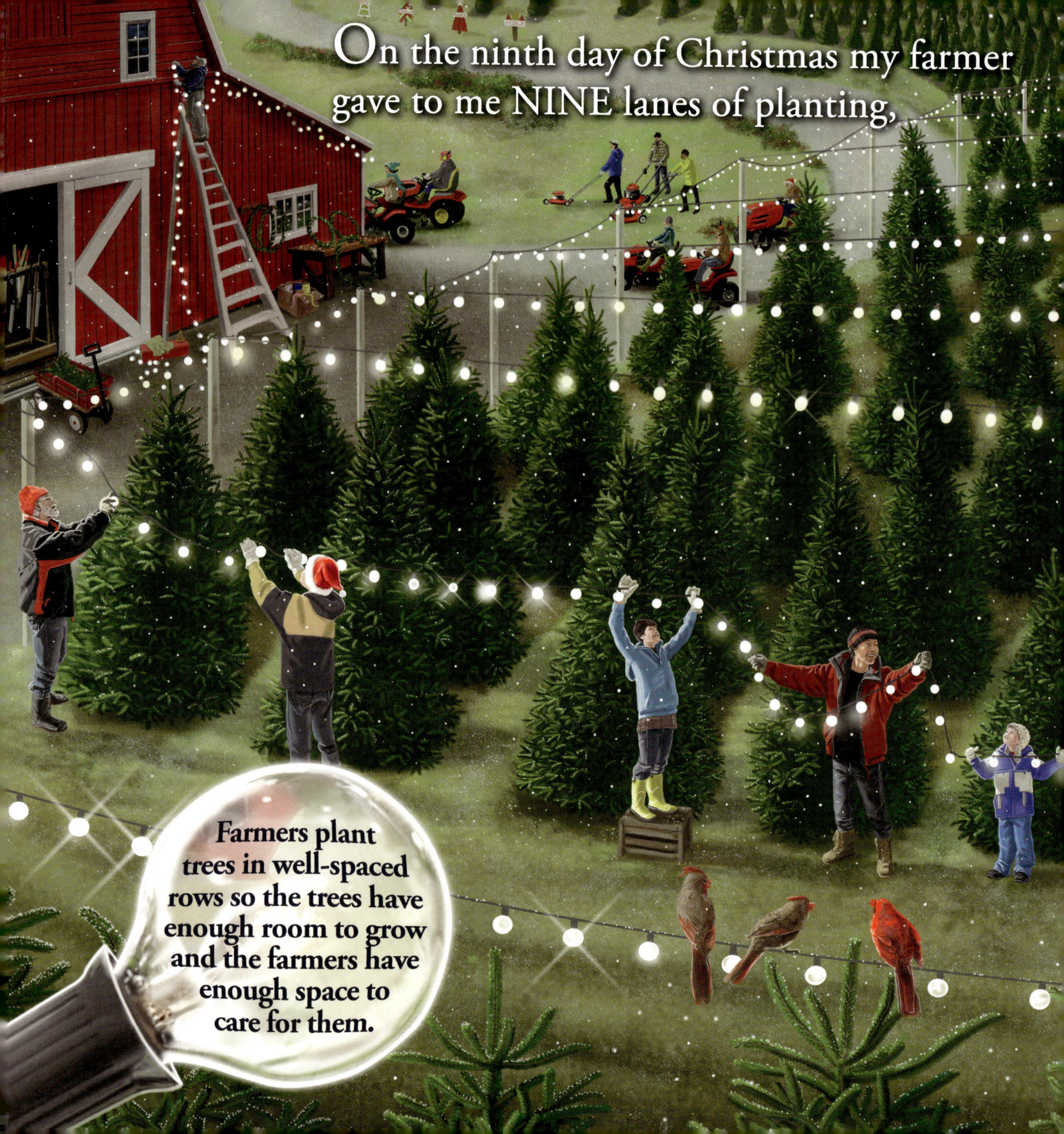

EIGHT mowers mowing, SEVEN shearers shearing,
SIX birds-a-nesting, FIVE woven rings,
FOUR kinds of Firs, THREE drenched friends,
TWO leather gloves, and a seed to grow a new tree.

On the tenth day of Christmas my farmer gave to me TEN hoards of seedlings,

A young tree is called a seedling. Christmas trees grow for about 3 years before they are planted at a farm. Farmers plant 1-to-3 new trees for each tree that is harvested.

NINE lanes of planting, EIGHT mowers mowing,
SEVEN shearers shearing, SIX birds-a-nesting,
FIVE woven rings, FOUR kinds of Firs,
THREE drenched friends, TWO leather gloves,
and a seed to grow a new tree.

On the eleventh day of Christmas my farmer gave to me ELEVEN taggers tagging,

Farmers often tag trees when they are ready to be harvested. Some farmers allow visitors to come tag their own trees before picking season.

TEN hoards of seedlings, NINE lanes of planting,
EIGHT mowers mowing, SEVEN shearers shearing,
SIX birds-a-nesting, FIVE woven rings,
FOUR kinds of Firs, THREE drenched friends,
TWO leather gloves, and a seed to grow a new tree.

On the twelfth day of Christmas my farmer gave to me TWELVE trees for trimming,

There are close to 350 million Christmas trees growing on farms in the US, all planted by farmers to help you celebrate the season!

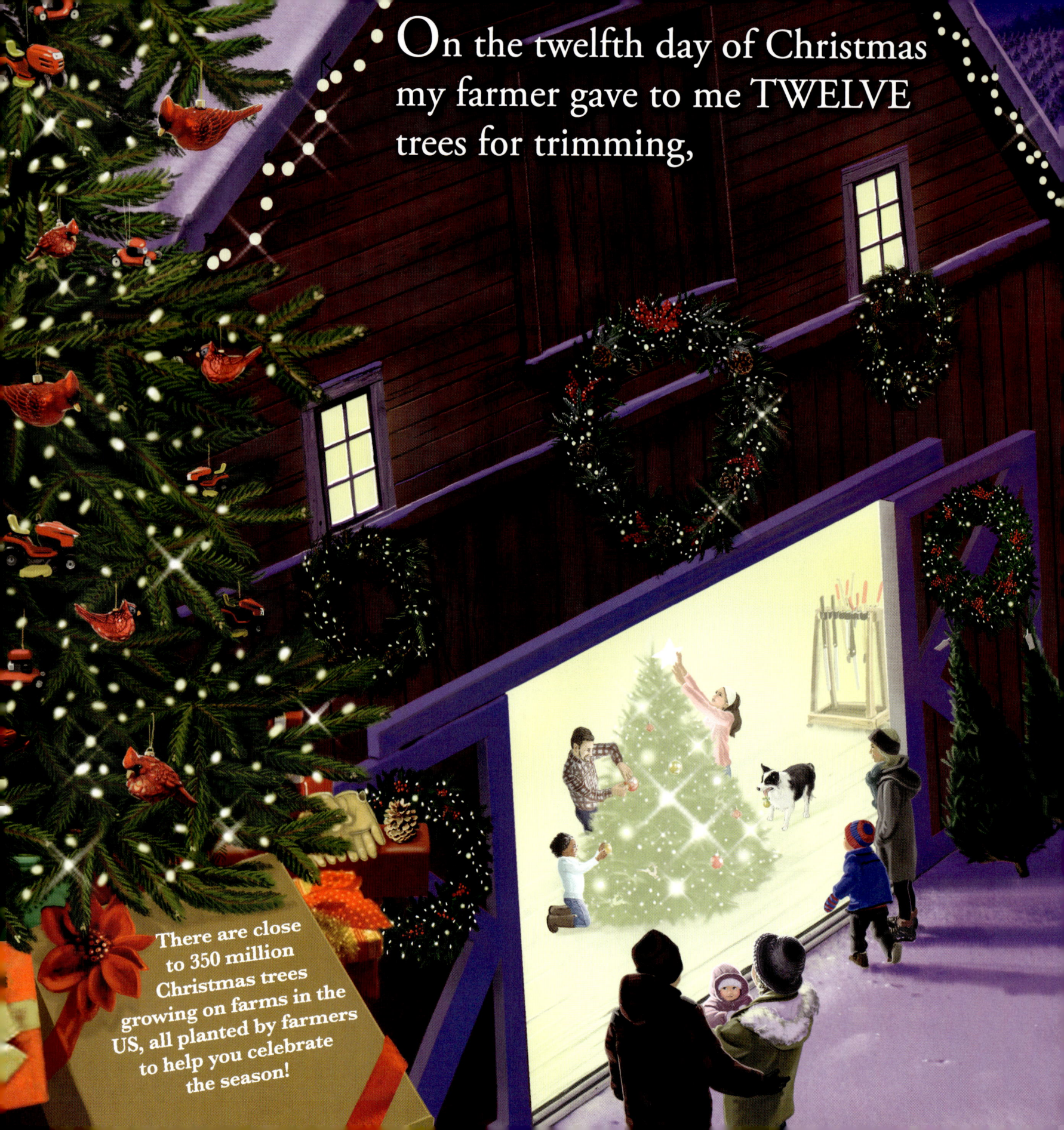

ELEVEN taggers tagging,
TEN hoards of seedlings,
NINE lanes of planting,
EIGHT mowers mowing,
SEVEN shearers shearing,
SIX birds-a-nesting,

FIVE woven rings,
FOUR kinds of Firs,
THREE drenched friends,
TWO leather gloves,
and a seed to grow a new tree.